THE MAMMOTH CAVE

AND ITS DENIZENS:

A Complete Descriptive Guide.

By

A.D. Binkerd, M.D.

Cincinnati:
Robert & Co., Printers
1869.

Table of Contents

To my Brother,

ISAAC B. BINKERD, ESQ.

in

Homage of his long and sincere devotion to the

CAUSE OF EDUCATION,

and in grateful recollection of his generous aid

and encouragement,

This little volume

is respectfully and affectionately inscribed

by

his former pupil,

the Author.

MAMMOTH CAVE, KENTUCKY.

LOCALITY.

The locality of the Mammoth Cave is in latitude 37° North, and longitude 9° West from Washington. Its only known entrance is in the eastern part of Edmondson county, Kentucky, on the south side of Green river, one hundred and ninety four feet above the level of that stream and ninety four miles nearly due south from Louisville.

MEANS OF ACCESS.

The Louisville and Nashville Railroad passes within a few miles of the cave. This road was projected as early as 1850, and the first through train passed over it on the 9th of November, 1859. It is now one of the best roads in the State, and a part of the great line extending from Chicago to New Orleans.

CAVE CITY,

eighty five miles south from Louisville and one hundred miles north from Nashville, is the point at which tourists stop over to visit the great Subterranean Wonder of the western world. A daily line of Concord coaches has long been established between this station and the cave.

Mr. W. L. Myers is the genial proprietor of the Cave City Hotel. He has, always on hand, something very nice expressly for you.

Mr. M'Coy, the proprietor of the stage line, keeps excellent stock, and employs none but the most careful and competent of drivers. The jaunt of nine miles over the country is a very pleasant one. Several points on the road command a wide range of beautiful scenery diversified by rocky hills and fertile plains.

RAILROAD PASSENGERS,

holding through tickets, may stop over, visit the cave, and resume their journey at pleasure on the same ticket. Many persons avail themselves of this privilege, generously extended by the railroad company to their patrons.

THE CAVE HOTEL.

This is rather a primitive edifice, constructed in the form of the letter L. It is, in the aggregate, over six hundred feet long, and has a wide, covered porch along the sides facing the inclosed angle. Fronting this promenade is a beautiful lawn, thickly shaded by natural forest trees and ornamental evergreens.

Mr. L. J. Proctor and Son are the present proprietors of the Cave Hotel and the Cave.

SOURCES OF AMUSEMENT.

A couple of billiard tables; a dancing hall, thirty by ninety feet, and a natural park of ambitious proportions afford the chief amusements of the place. In the months of May and June the air is fragrant with the aroma of roses and other flowers cultivated in the garden adjoining the buildings. On fair days the wild birds beguile the hours with their varied songs. During the traveling season a band is employed to discourse music to the patrons of the dance.

ENTRANCE TO THE CAVE.

A few minutes walk, out through the garden, over the stile, and down a flight of wooden steps, brings us into a rocky ravine deeply shaded by tall forest trees. Here the air is cool and bracing. The sensation is delightful, and we catch new inspiration from each long, deep draught of the vitalizing element.

Proceeding on our way, we presently reach a dilapidated old log building, in front of which there is a yawning chasm fifty feet deep, with irregular and precipitous sides. This is the dreary portal to the subterranean world. Green ferns and climbing vines cling everywhere to the projecting rocks as if striving to cast some adorning drapery about their nakedness.

THE CASCADE.

A little spring of water pours a ceaseless stream of silvery beads from a shelving rock above the entrance and dashes it to spray in the chasm below. One fancies that the monotonous hum of the falling water and the gloom of the thick, overhanging foliage, render the place a fit habitation for gnomes. The first emotions awakened at sight of the entrance, and its weird surroundings, are less agreeable than we could have wished.

THE OLD ENTRANCE.

Formerly, ingress was effected farther down the hill, near the Green river, where the cave may still be entered and explored as far as the breach forming the present entrance. At the old entrance we walk into the cave on a horizontal line, as into a coal mine or railroad tunnel.

That part of the cave between the old and the new entrance is about half a mile long, and is known as Dickson's Cave. It contains nothing of special interest and is rarely visited.

THE NEW ENTRANCE.

At the new entrance we descend into a deep pit or shaft till we reach the floor of the cavern, about on a level with the old entrance. The present entrance was occasioned probably by the action of a little stream of water, causing the rocky roof, which was not very firm at this place, to break through. A knowledge of this fact may excite apprehensions of danger, but having once entered the cave, a sense of security steals over us, and we dismiss fear.

TEMPERATURE.

In these rocky chambers the temperature is uniformly about 59° F. The cave exhales or inhales as the temperature outside is above or below this standard. In summer, a strong current of cool air rushes outward with such violence as to endanger our lights. In the cold weather of winter, the current sets inward. In the spring and fall, when the temperature outside is about equal to the temperature inside of the cave, there is no action whatever. This natural phenomenon is called the breathing of the cave.

Coming out of the cave on the last day of March of this year, I noticed a strong current of air tending outward, increasing in violence as I approached the entrance (it being a fine day).

When I had approached so near to the door as to be able to recognize the gray dawn of the daylight without, I tried the effect of this current of air upon my lamp, and found it sufficient to extinguish the flame. I had carried the lamp more than two hours, and it was partly exhausted. The breathing inside of the cave is never perceptible more than a few hundred yards from the entrance, except in case of a violent storm raging without, accompanied by sudden and great change of temperature.

Change of season is unknown in the cave. Mornings and evenings have no existence in this nether world. Time itself produces no change in many parts of the cave; for where there is no variation of temperature, no water and no light, the rocks may defy the three great forces of geological transformation.

EXPLORING PARTIES.

Exploring parties are not properly equipped for the underground journey until each member of the company is provided with a pair of thick soled shoes or boots, a cap, blouse, and staff. The ladies should be arrayed in Turkish costume, with a hood of woolen stuff covering the head and ears.

THE DESCENT.

The guide, with a canteen of oil slung to his side, a box of matches and a good supply of Bengal lights in his pocket, and a basket of refreshments on his arm, hands to each a lighted lamp; then leads the way while we follow down a flight of rude stone steps till we reach the floor of the cavern. Here we pause a moment, take another look at the sunny sky, and then pass behind the sheet of falling water and enter the door in the artificial wall that separates the outer world of sunshine from the realm of darkness. From the time we pass this door, our minds are so occupied with new and interesting sights that we rarely think of anything else till we return to daylight again.

FIRST IMPRESSIONS.

On entering the cave we feel a slight chilliness, and perhaps, too, a touch of fear; but these sensations soon vanish as gently and imperceptibly as childhood glides into youth. Before we have gone half a mile we feel ourselves the dauntless explorers, of unknown realms, ready to enter the darkest regions of the cave, guided by the light of a single taper. The courage of the timid tourist sometimes wavers before reaching the cascade, but we never knew any one to turn back voluntarily after having gone as far as the vestibule.

We can not see distinctly for some time after entering the cave. But by the time we shall have reached the first point of considerable interest, the eye will be somewhat accustomed to the darkness, which will enable us to see more clearly.

THE NARROWS.

For the distance of fifty yards or more beyond the entrance, there is a low narrow passage with an artificial wall on each side, rudely constructed of the fragments of rock that were quarried from the bottom and forced from the low ceiling, in order to enable a span of oxen and a cart to enter the cave. These were employed in the manufacture of saltpetre or nitrate of potash, which was extensively collected here from 1808 to 1814, by persons in the employ of the United States Government. The numerous rude appliances that were used in the manufacture of this salt, are still scattered along in the cave. Many articles of wood and some of iron may still be found here as firm and fit for use as when they were laid aside over half a century ago.

FIRST VATS.

Just beyond the Narrows, on the right hand side, are two huge bins or boxes, twelve feet long, six feet wide, and four feet deep. These boxes were constructed of strong oaken plank, and are still full of leached dirt that is almost as firm as a block of limestone. These bins, with their contents, have been carefully preserved, just as they were left by the miners in 1812. A great number of the trunks of thrifty young poplar trees, from eight to twelve inches in diameter and from twenty to twenty five feet long, perforated longitudinally with a two inch auger, lie scattered along the floor, from the entrance to the distance of half a mile into the cave. Through one line of these old pump logs, fresh water was conducted from without for the purpose of leaching the dirt, and through the other the lixivium was forced back by means of a hand pump, to the entrance, where it was evaporated to crystals.

THE EARTHQUAKE OF 1812.

Mr. J. Gatewood, a native of the county, and an employee in the saltpetre mines, frequently stated during his lifetime, within the hearing of Mr. A. L. Mallory, my informant, that he was in the cave with a number of other workmen, during the occurrence of the earthquake that formed the lake in the lowland known as the "Kentucky Purchase," in the southwestern part of the State, bordering on the Ohio. The tremulous motion of the earth filled the miners with alarm, and they fled in the wildest confusion toward the entrance, which they did not reach till long after the danger was past, when they stepped forth with thankful hearts from what they feared might prove their sepulcher. Fortunately no one was hurt, nor were the mining operations interfered with. Since the cave has proved a safe retreat during a violent earthquake, it is hardly probable that it could be unsafe at other times. No accident or loss of life has yet occurred in the cave, from carelessness or foul play, within the memory of that reverend being, the oldest inhabitant.

THE ROTUNDA.

We next enter the vestibule or rotunda. This is a large cavern at the beginning of the main cave, and is said to be directly under the hotel. It is over seventy five feet high, and one hundred and sixty feet across the floor. Several avenues put off in different directions from this, as from most other of the large rooms. In some parts the wall is abrupt, in others the ceiling slopes down gradually to the floor. Off to the right is a passage rarely shown to visitors, as it contains nothing of special importance. This is called

AUDUBON'S AVENUE.

Being near the entrance, it is generally passed by without much attention, and the guides have finally dropped it from their course altogether. One part of this avenue presents an unusual attraction to the naturalist. Countless thousands of bats have taken possession of one of these caverns, wherefore it is known as

THE GREAT BAT ROOM.

Here the leathern winged little animals cling to the walls and ceiling like huge swarms of bees, in bunches of many bushels, and doze away their existence in a semi-torpid state, in darkness and repose. What they feed upon is a question not easily settled. Some of them must remain here many months without once going out of the cave, as but few are seen outside at a time. It was formerly believed that they spent only the winter here; but I have never visited their apartment at any season without, finding numbers of them. We will find plenty of them in the Rotunda for any experiment we may wish to make.

They are cold to the touch, and when seized between the thumb and finger, they shrug up their shoulders, move their wings lazily, and perhaps utter a feeble cry. Their eyes are about the size of a cambric needle's head; these they persistently close when brought to the light. I carried one out of the cave, and after examining it to my satisfaction, threw it up into the air, thinking it would fly, but it only used its wings as a parachute, lit gently upon the ground, and did not seem to have vitality enough to appreciate its freedom. They are very small, not over an inch and a half, or, at most, two inches in length, measuring about six inches across the wings.

Notwitstanding their diminutive size, they are nevertheless capable of inflicting a severe wound. If we hold one of them by the fur and skin on the back of the neck, and blow into its face, it will curl up its nose as if in derision, open wide its mouth and display a beautiful set of the most perfect and delicate teeth, similar to a cat's and sharp as a needle. These little animals are classed by naturalists with the true quadrupeds or mammals, as they bring forth their young alive; and are grouped with the carnivora or insectivora.

THE MAIN CAVE.

The main cave begins at the Rotunda, under the hotel, and extends to the distance of five or six miles. It varies in width from fifty to three hundred feet, and in some places it is one hundred feet high. For the distance of a mile it is straight, then turning to the left it forms with itself an acute angle; after which its course is irregular. Some of the small passages putting off from the main cave, after prolonged and tortuous windings, communicate with other caverns and domes, surpassing in grandeur and magnificence even the most renowned part of the main cave. We cannot, in a small work like this, give a minute description of all that is interesting in so great a cavernous region. We propose to mention in the order of their occurrence only the main points of interest to which the tourist's attention is called, stating such facts regarding them as we have been enabled to gather by diligent research, inquiry of persons now living, and by personal inspection.

KENTUCKY CLIFFS.

As we move forward in the main cave, we notice on our left a rough ledge of beetling rocks, resembling the cliffs on the Kentucky river, after which they are named. On the right there is nothing peculiar, save sometimes a gradual sloping of the roof toward the floor.

THE PIGEON BOXES.

About four feet up the left hand wall there is a cluster of holes, very regularly formed, and about large enough to admit the hand. These being a distinct group, limited in number, while the adjoining wall is smooth, naturally suggests the name of Pigeon Boxes.

THE CHURCH.

About a quarter of a mile beyond the Rotunda, we enter a second dome or enlargement, in the main cave. This has a gothic roof or ceiling spanning the vast arch, forty feet above the floor. The hall is somewhat irregular, and has an area of many thousand square feet. At the left hand corner as we enter this hall, there is a solid stone projection or platform about three feet higher than the main floor, and wide enough to hold a stand and several chairs. This is called the pulpit, and from it the Gospel was formerly preached to the large and attentive audiences that were probably attracted thither by the novelty of the occasion. These old pump logs arranged into rows of seats may still bear testimony that the story of Christ crucified has been told even in the sunless caverns that underlie the "dark and bloody ground." A rude gallery extends around a part of this hall, perhaps twenty feet above the main floor.

SECOND HOPPERS.

We have now reached a second series of vats or shallow pits, constructed of round sticks or split logs. Some of these are full of dirt and others are empty, resembling old pig pens. The great number of these rude appliances still remaining give some idea of the extent to which the work of saltpetre mining was carried on here. Notwithstanding the imperfect state of chemical knowledge half a century ago, and the primitive method employed in extracting the salt, it is said the yield of a single year was estimated in value at twenty thousand dollars. These mines are very rich, and it is believed that the dirt which has once been leached, has the power of absorbing this salt again from its great source, and may be worked over with profit every three or four years, thus affording an endless supply of the nitrate of potash.

GOTHIC ARCADE.

Just in this part of the cave, where the mining operations were most extensively carried on, we notice a flight of wooden steps leading up to a large opening in the wall on our right. This is the entrance to a very interesting part of the cave, which we will visit on our return. As it is the best part of the day's work, we will reserve it to the last, make dessert of it.

THE BALL ROOM.

The Ball Room is the next place of interest that claims our attention. It is an enlarged portion of the main cave, perhaps one hundred feet long, sixty feet wide,and forty or fifty feet high. The floor is even, the walls and ceiling are regular, and with a little labor it could he made one of the most charming and commodious halls in the entire series. But the proprietors are anxious to show the cave as nearly in the natural state as possible, and therefore all the embellishments of art are scrupulously discarded.

OX TRACKS IN THE ROCK.

In this part of the cave are still to be seen the tracks of cattle and of the carts that were used by the miners. In one place is a distinct ox track in a hard substance similar to limestone rock. It will be borne in mind that these tracks were made in the soft mud over half a century ago, and since this part of the cave is very dry, and has been so for many years, the mud has become so thoroughly baked that it has assumed the consistency of rock. Thousands of human feet have trodden over it, and still the intaglio remains indelibly fixed in the hard substance. Cart tracks are too numerous and well defined to be mistaken by even a blind man.

OLD CORN COBS.

Close by the wall on the left side of the room, may still be seen the fragments of corn cobs, but whether they were brought here by the miners, as is the tradition, or by Mat, the guide, for the purpose, as he says, of gratifying the curiosity of relic hunters, we have no means of determining. Three or four feet up the wall from these fragments, where the cattle were most probably fed, there is a kind of hitching ring naturally formed in the rock. To this the guide informed us the cattle were fastened, during the intervals allowed for rest and for taking nourishment. The smooth groove worn by the rope into the rock, proves beyond doubt that the prevailing opinion has some foundation in truth.

THE STANDING ROCKS.

At the farther end of the Ball Room, there are several large flat rocks that must have tumbled from the ceiling, perhaps ages ago. Several of these most probably turned edge foremost in the descent, in which position they buried themselves so firmly in the dirt and rubbish, that they still stand as unyielding as the Leaning Tower of Pisa. Two of these deserve special attention. They are, perhaps, fifteen feet long and project ten feet above the floor. The one nearest the wall stands nearly perpendicular, and has a heavy molding along the top. The other is about two feet thick, and stands parallel with the first, three feet from it, but leaning over toward it at the top. Playful boys and girls seldom fail to pass between the standing rocks. There are no dangerous pits in this part of the cave. The floor is covered with a fine dry dust that never rises like the dust of the outer world, in spite of the pattering of many feet. It will neither adhere to a polished boot nor scarcely soil our garments.

GRAND ARCH.

This portion of the cave is very similar to that which we have already seen, except the peculiar arch in the ceiling, which, when well lit up is one of the grandest sights we have yet witnessed. Like a painting, it should be studied in order to get the best effect. It will amply repay us for all the time we spend in contemplating its beauty. This arch is about fifty feet high and sixty feet wide.

WILLIE'S SPRING.

On the left side of this beautiful stretch of wonders, a thread of water about the size of a Faber pencil, has chiseled a fantastic little niche into the solid limestone rock, and gathered itself into a spring or basin at the foot of the niche. Tasting the water, we find it fresh and palatable. A little reflection here, upon cause and effect, will do much toward explaining the theory of cave formation. Three conditions are required to enable the water to carve these magnificent halls into the solid rock. These conditions all exist at the spring above referred to. The limestone is a very pure carbonate, the water contains carbonic acid, and in order to do work it must be in motion. The water containing the carbonic acid, in running over the rock, dissolves, takes up and holds in solution the lime and carries it away. While this process is going slowly on, at a very low temperature, the conditions are every instant renewed, by which the work must be continued. While the water is taking up the particles of lime, the carbonic acid is liberated from its union with the lime in the rock, and now the acid is taken up by the water to which it adds the solvent power over the rocks. The peculiar shapes in the various caverns are due mostly to two causes: first, the different degrees of solubility in the different parts of the same rock; and, second, the current or motion of the water. The tourist will observe many freaks of shape and direction in niches and caverns directly traceable to the causes above enumerated.

THE TIMEPIECE.

Behind some loose rocks on the left may be heard the slow dropping of water, at such regular intervals as to imitate the ticking of a timepiece. It requires but little effort to imagine oneself in a quiet old fashioned house, lulled to repose by the drowsy tickings of the old moon faced clock. Mr. Samuel Meredith, the guide, informed us that this natural chronometer continues to mark time, from year to year, with wonderful uniformity of intervals between drops.

ROCKY HALL.

On our left there is a high opening in the wall half blocked up with huge boulders or immense rocks that have tumbled down from time to time. Notwithstanding the roughness of the passage, it may be followed for more than two miles, but it is doubtful whether we would feel sufficiently rewarded for the trouble should we attempt to explore it. It is never shown to visitors.

PICTURES ON THE WALL.

Now we are upon the threshold of that part of the cave so full of wonders and attractions for all lovers of pictures. The prevailing color of the floor, walls, and ceiling of the cave is dark gray. Here the ceiling is covered with an incrustation of gypsum, manganese or iron, producing the different shades of white or dark in proportion as the gypsum or the iron predominates. In some cases there is a dark picture upon a white background, in others a white picture upon a dark background, and in still others, a black and white picture upon a gray background, giving the whole a variegated and fanciful appearance. We pause here a minute and look around, tracing out the different kinds of animals. In the dim light of our lamps, they appear wonderfully true to life. The longer we look the more difficult it becomes to resist the impression that we have actually entered the studio of some young artist who has just blocked out a first class menagerie.

Here is a pair of pretty good bears; there, a group of monkeys; yonder, a wildcat; farther over, a veritable elephant; and farther on is a catamount crouching upon a log ready to spring upon its victim. See how it gazes with an expression of fierce earnestness that might defy the king of beasts! Over to the left is the big Indian, tossing his papoose playfully over to his squaw, seated at his feet. Many of these pictures are in various attitudes, displaying a grace of motion as nimble and airy as if they had been photographed upon the wall, by some magic art, with the suddenness of a flash of lightning.

THE GIANT'S COFFIN.

Just one mile from the entrance, on the right side of our path, there is a large rock, fifty seven feet long, detached from the rest, and standing up a little from the floor. This bears so perfect a resemblance to a huge coffin, that any one can see the fitness of the name of "Giant's Coffin," by which it is known. On our left the wall is abrupt, and the ceiling high above our beads, but on the right and in front the roof gradually slopes down toward the floor. Here the cave makes an acute angle with itself, and just at the apex of the angle is

McPHERSON'S MONUMENT.

This is a rude pile of unhewn stone, erected by his surviving staff officers to the memory of gallant soldier whose name it bears. A stone is occasionally added to the pile by one of the

General's friends, but like most other things in the cave, its growth is rather slow.

Here we will leave the main cave awhile for scenes of a different character. Our path leads around the Giant's Coffin to the right. Now we must stoop, in order to enter a low, tortuous passage that leads downward and perhaps backward under the main cave to a series of rooms of considerable size known as the

DESERTED CHAMBERS.

The direction of these caverns is by no means regular, nor do they continue on the same plane. The general tendency of the grade of the main cave is downward from the entrance, and the same is true of the passage we are now following. When we shall have reached the rivers, about a mile from this point, we will have descended nearly to the level of the Green river, which is, as already stated, one hundred and ninety four feet below the entrance to the cave.

STEPS OF TIME.

Here we descend a flight of ten or twelve wooden steps, marvelously steep; then turning a little to the right, we follow a more convenient path that leads to a large room, one hundred feet in diameter. The floor is irregular and the ceiling low and concave. This is called

THE WOODEN BOWL.

It is probable that even this part of the cave was known to the Indians, for it is said that a wooden bowl of rude construction was found in this room, by the first white man that explored it. Here the wonderful formations begin to crowd on us thick and fast, and many objects of rare beauty will probably be passed unnoticed.

BLACKSNAKE AVENUE

leads from Wooden Bowl to the main cave. It derives its name from its serpentine course and black walls. It is now rarely visited.

MARTHA'S PALACE.

This is a conical little opening, dissolved into the solid rock that forms the roof above our heads. In all this region the slow work of rock carving is still going on. The little drops of water, falling at such intervals that they may be counted, are the assiduous agents by which these magnificent halls are fashioned.

RICHARDSON'S SPRING.

The water does not only drop from above, but it also wells up from below; for here in the midst of our path is a little basin of it, ready to quench the thirst of every corner. See how it sparkles as it runs away to resume its carving, humming a low song as it goes! The smooth rocks that may answer the purpose of seats, and that cup by the spring, tell plainly enough what the guide means by leaving the basket here. We shall dine here today as we return from over the Styx.

ARCHED WAY.

We continue our journey along a grotto of sufficient height to allow us to walk erect. The ceiling overhead forms a gothic arch, similar to that in the main cave. Though not so grand and imposing, it is of similar formation, but on a smaller scale. The character of this hall changes but little till we reach

SIDE-SADDLE PIT.

This is an irregular opening in the plane of the floor, bearing some resemblance to a side saddle viewed from above. It is forty five feet deep. Immediately over it is a huge irregular opening extending upward to the distance of forty feet or more. This is called

MINERVA'S DOME.

These immense caverns extend downward and upward from the level of our path. They are so close to the right hand side that we may view them thoroughly without any danger of tumbling into them. Due care, however, should always be taken to avoid accident. Just on the left, and a little forward, from this place, is the famous

SOUNDING ROCK.

This is a petrous formation nearly detached from the wall, resembling in shape the human ear. When struck by the hand it gives a low, sweet, musical tone. This point is rarely passed unnoticed. Close to the left side of our path is the mouth of the

BOTTOMLESS PIT.

This strange place was disarmed of some of its terror when it was ascertained that it has a bottom said to be one hundred and seventy five feet below the path, along which there is an iron railing to prevent sinners from straying into that pit. So far no one has ever come to grief in this yawning chasm. Immediately over the pit is a high arch, similar to that over Side-Saddle Pit. This is known as

SHELBY'S DOME.

Some very singular and beautiful formations may be seen up in this dome, by throwing a Bengal light in such a position as to illuminate it to the top. Scroll work, corrugated panels and fanciful projections are lavished everywhere in such profusion as to bewilder the eye. The click of dropping water is heard at measured intervals in the dreary darkness around, reminding one of dropping nuts, or the falling of autumnal leaves, when the slowly rising morning sun unlocks the icy fetters that had bound them to the parent stem. But we can not pause to conjure up reflections.

BRIDGE OF SIGHS.

This is a wooden structure three or four feet wide, and ten feet long, spanning a chasm between Side-Saddle Pit on our right and Bottomless Pit on our left. There is a wooden handrail on each side of this narrow bridge, so that there is no danger in passing over it. Just here there is an old ghost of a bridge that was thrown over an arm of Side-Saddle Pit, and then abandoned half finished, for fear it might decay, being kept constantly wet by the dropping water. No one ventures upon it.

WINDING LABYRINTH.

Just beyond the little wooden bridge we turn off abruptly to the right, and descend a long, steep wooden stairway, to the bottom of a narrow, winding gorge, scarcely wide enough for two persons to pass. We follow this winding gorge toward almost every cardinal point of the compass, ascend a second stairway, and then descend a third, and finally reach a window like opening on our left. Here we are upon the threshold of great wonders and indescribable beauties. This is

GORIN'S DOME,

and the crowning attraction of the day's journey. By holding our lights in through this opening, we see a huge curtain of limestone a hundred feet long, suspended far above and stretching its wavy folds of petrous drapery many yards below our feet, terminating about twenty feet above the floor. This dome is two hundred feet high and sixty feet across the widest part. The guide goes up to another opening and throws down a Bengal light, which enables us to take the dimensions of this vast cavern. Looking up we can see a huge glimmering dome a hundred feet above our heads. We catch but an imperfect glimpse of this great wonder, yet we are impressed with its grandeur and matchless beauty. But we can not tarry here, so we retrace our steps toward the arched way.

THE AMERICAN EAGLE.

As we return through this narrow gorge, just after descending the flight of wooden steps, we see on the rocky wall to our right, a huge spread eagle, fixed immovably upon the rock. It is not the work of art, but of Nature's own fashioning; perhaps not one of her best specimens, but still it can easily be recognized. Having ascended the long flight of steps, we find ourselves in the same grotto we started from.

REVELERS' HALL.

This is a wild, triangular room, rough and irregular in outline. Besides the passage by which we entered, there is one leading off a little to the right, and quite large enough to admit a man on horseback. It may be followed for a mile or more, and contains some objects of interest. It is not shown to visitors except on rare occasions. The chief attractions are Pensacola Avenue, Wild Hall, Snowball Arch and Pine Apple Tree.

SCOTCHMAN'S TRAP.

By some frightful commotion among the rocks, perhaps ages ago, the most elaborately ornamented parts of the cave came near being veiled from human sight forever. The huge stone cover, that would have blocked up this passage to the rivers, but for the apex of its angle, that caught against the opposite wall, still leans over the door, at an angle of forty five degrees. With some misgivings we pass under this trap, descend a declivity, and enter the

VALLEY OF HUMILITY.

Here we perform a grievous amount of stooping, and but for the name of the passage, which serves to keep us in a serene and amiable mood, we might recall some of the characters in Dante's inferno.

Suddenly we enter a new passage at right angles. This is a little higher in the ceiling, and allows us to breathe freely. The part leading to the right is called

BUNYAN'S WAY.

This may be followed for miles of tortuous windings that only perplex the explorer by bringing him into his old path again. That leading to our left is called

BUCHANAN'S WAY.

This may also be followed for a great distance without coming to anything very definite or specially interesting. We choose to follow it, however, for about forty yards. Here we will turn of at right angles into that unique passage known as the Winding Way, or more familiarly as

FAT MAN'S MISERY.

This is a narrow, meandering path, one hundred and five yards long, cut or dissolved by water, about eighteen or twenty inches wide, and three feet deep into a solid rock, and of wonderful uniformity in both width and depth. This serpentine channel bears the above name because it will not permit any one to pass, whose size exceeds certain prescribed limits. The largest man who was ever known to pass here, weighed two hundred and sixty pounds. It was not without many a labored effort that he succeeded in emerging from its inner portal. Before this channel was cut, there seems to have been a horizontal aperture of several feet between the different strata of limestone, and extending to a considerable distance on each side of the lowest part naturally sought by the water. From the bottom of the channel to the ceiling, the distance varies from four and a half to six feet. The path is a perfect negative or lithograph of running water. The little waves and ripples stand out from the sides and bottom in bass relief, as if the water had been suddenly converted into stone.

After being thoroughly ground in this relentless mill, we emerge into a large open hall. Here we are glad to take a breathing spell, and devise some way, if possible, by which to get back without doing penance a second time, for possessing a

30

full stature. But there is no escape from a second trial, unless we choose to remain in the realms of eternal darkness. The fates are as unrelenting as Neptune, when the voyager is about to cross the line for the first time.

GREAT RELIEF.

This is a cavern of considerable size, entered at right angles by the winding way. We will follow it to the right. The same characteristics abound here as elsewhere. No odor of any kind is perceptible. The temperature is the same as near the entrance. There is no motion whatever in the air. No signs of animal or vegetable life are met with, except perhaps a few rat tracks. An almost palpable darkness and a painful stillness pervade this gloomy region perpetually, except when the lamps of an exploring party shed their feeble light in these grand halls, or the swollen rivers dash against the rocks with sullen roar.

ODD FELLOWS' LINKS.

Three large links of a powerful chain are stretched across beneath, and firmly adherent to the ceiling.

They are each about two feet long and as thick as a stout man's arm. They are somewhat discolored, showing the presence of a trace of iron in the limestone composing them. The peculiar shape and disposition of the insoluble limestone are among

31

those freaks of nature we constantly meet with and yet are unable to explain. Just a few rods farther on are two prominent seams or ridges, the one branching off from the other, and resembling somewhat in shape and direction in relation to each other a portion of the junction of the

OHIO AND MISSISSIPPI RIVERS.

These prominent ridges, like the links, contain a trace of iron which resists the action of the water, while the more soluble limestone is dissolved away. They are fifteen or twenty feet long, and stand out in bass relief several inches from the ceiling.

RIVER HALL.

We have now reached the point to which the water sometimes rises in the cave during a very wet season. Low water mark is forty feet below this point. On the first of April of this year, the arches over the river were only a few feet under water. A fall of six feet would make the navigation of the rivers and the exploration of the entire cave quite practicable.

The statement that these rivers are subject to sudden tidal fluctuations has engendered some prejudice in the minds of the people against crossing them. It is true they do rise and fall, but this is due to neither lunar nor mysterious influence. They rise as the Green river rises, with which they doubtless communicate by unknown passages. They fall much slower than the Green river, because these passages are too small to permit the water to escape as rapidly as does that in the channel of Green river.

But, says one, how shall I understand that? Will more water run into a barrel than will run out of it in the same time, the apertures of ingress and egress and the pressure being the same? Certainly not! The same cause that will make the Green river rise, will also make the fountains of the regions round about rise, and why not the rivers in the cave? These rivers do not rise suddenly without some obvious reason for it any sooner than other rivers.

The general tendency of the rivers in the cave is that of subsidence. The Lethe, which was a running stream a few years ago, is now a standing pool at low water. The waters have cut deeper channels and sought lower levels. The caverns we now navigate in boats will probably some day be passed through by the tourist on dry shod, unless the level of the Green river should arrest the descent of these streams.

THE DEAD SEA.

This dismal looking pool of standing water is in the hall above described. The surface is from forty to fifty feet below the floor of the hall, from which it is viewed. The surface of the Dead Sea comprises an area of about a thousand square feet. If we throw a stone into it as almost every one is tempted to do, there comes back a thud that seems to awaken a legion of slumbering echoes. The water of the Dead Sea is perhaps twenty feet deep. At one place we may approach very near its surface and almost feel the strange gloom that hovers over it.

THE RIVER STYX.

The source and mouth of this mystical river are unknown. It is probably in some way connected with other rivers in the cave, as it rises and falls at the same time with them. It is thirty or forty feet wide, and varies from twenty to fifty feet in depth, and is five hundred feet long. It may be navigated the whole length of the stream.

THE NATURAL BRIDGE.

This is a limestone arch spanning the Styx twenty five feet above that stream. We will pass over this bridge. It is well to be careful in this region of the cave, as there are several slippery places and low ceilings. With ordinary caution no harm can befall any one; yet a misstep might subject one to an unpleasant bath in the Styx. The next point of interest is

LAKE LETHE.

A strange feeling creeps over us as we look for the first time upon this quiet pool of water and the little boat moored to its recently discovered shore. On each side is a wall of massive rock, and overhead an arched ceiling spans the chasm ninety feet above the water. We are now about three hundred and seventy five feet beneath the surface of the earth, beyond the reach of light or heat, burn the sun never so fiercely. No sign of life is here, save what we bring and will soon take away again. No carking cares pursue seething, fretting humanity to these dreary halls. We have forgotten the outer world, absorbed in transports of awe and admiration.

This is the end of our explorations in this direction today. Tomorrow we will begin at this point the far more interesting work of navigating the rivers and exploring halls of matchless brilliancy, more charming in scenic effect than oriental palaces. We will now retrace our steps over the Styx by the bridge and examine for a few minutes

THE BACON CHAMBER.

This is a low cavern in one of the grottoes putting of from River Hall. It is mostly interesting for the peculiar formation of its ceiling. The pendant masses of limestone rock bear a striking resemblance to hams of meat. A little to one side there

35

is wrought into the ceiling a huge inverted kettle or large bell without a clapper. These fantastic formations were all carved into the solid rock by the action of water. But how the water was enabled to act against gravity is not so clear to my mind. One theory, however, presents itself as a solution to the question. This ornamented ceiling is just a little above the present high water mark, and as there is a vast deposit of sand upon the floor, it is natural to suppose that the water rose periodically high enough to touch the ceiling, and by continued waving motions succeeded in carving out these formations, resembling the contents of a smoke house.

As our allotted time is about expired, we will hasten back, thread again the Winding Way, probably under protest; fight our stooping battles over again; emerge from beneath the Scotchman's Trap, and quench our thirst at Richardson's Spring. The guide now brings forth the basket, whose savory contents are eagerly appropriated. The sandwiches and cold chicken vanish from sight as the frost before the rays of the morning sun. Our physical forces are renewed, and we inwardly thank

> "The powers that make mankind their care,
> And dish them out their bill of fare."

THE ATMOSPHERE

is a mild nitrous oxide, so cool and bracing that we may journey for miles without feeling exhausted. The proportions of oxygen and nitrogen bear the same relation to each other in the atmosphere of the cave as they do in the air of the outer world, but the average of many observations shows the carbonic acid gas to be much less. In the dry parts of the cave there are about two parts of carbonic acid gas to ten thousand parts of air. In the vicinity of the rivers there is still less of the former. In those parts not commonly visited and not frequented by bats, no ammonia has been detected, nor have the most delicate reagents shown the slightest trace of ozone.

The amount of moisture varies in different parts of the cave. The hygroscopic properties of the nitrate of lime that is found in many of the rooms prevent the putrefactive decomposition of all animal matter. Defunct animals will mummify rather than decompose. The bodies of two Indians were found in the cave many years ago. They were dry and well preserved.

Delicate females frequently do the long journey. It is their almost universal testimony that they have experienced more fatigue from a single hour's shopping than from this underground journey of eighteen miles. Some languor is usually felt after coming into the warmer air of the outer world, but the refreshing sleep usually enjoyed on the following night exercises a powerfully restorative influence.

ON THE RETURN.

As we have two points of great interest still before us, we press onward, pausing occasionally to inspect some points of interest too hastily passed before. In a short time we again enter the main cave, behind the Giant's Coffin. We can not recognize the peculiar shape of this rock till we come round in front to the point from which we first viewed its singular proportions.

THE ACUTE ANGLE.

This is the point at which we left the main cave a few hours ago, and just here we will take up the clue and pursue our observations in this high hall. The same architectural analogues are apparent at every step, and the ceiling is covered with grotesque pictures of men and beasts. Our eyes have become so accustomed to the pitchy darkness that we see even better now than at any time before.

THE DESERTED VILLAGE.

Just beyond the angle of the main cave, there are still standing the roofless remains of several stone cabins, which were erected over a quarter of a century ago. These were occupied by a number of consumptives, who for a time took up their quiet abode here, hoping to be benefited by a permanent residence in the cave. I believe there were fifteen of them. They entered in September, 1843, and remained till January. Three of the number died here, and the rest not feeling materially benefited, left the cave, and nearly, if not quite all of them, died soon after.

EFFECTS UPON ANIMAL AND VEGETABLE LIFE.

As a sanitarium, the cave will probably never be a success. Since the sun is the great source of light, heat, and therefore life, it is hardly probable that the protracted absence of its influence can be of any therapeutic value in chronic diseases. The contrary may be expected, except in some cerebral affections.

A number of trees of various kinds were planted near these cabins, and notwithstanding they were carefully irrigated, they showed no more signs of budding than if they had been stone or iron. The first time we entered the cave, the sapless remains of some of the trees were still standing as so many witnesses, bearing testimony to the fact that the light of the sun is as necessary to the life, health and growth of plints and trees as moisture or warmth. The absence of any one of these three conditions for a considerable time, is equally fatal to vegetable life. No experiments have yet been made to ascertain the effect of this intense darkness and strange stillness upon persons in health.

It is said that the cattle that were used in the cave grew sleek and fat in a few months, without extra feed. Persons in ordinary health, or even invalids, stimulated by the novelty of new and beautiful sights at every advance, may remain in the cave for a few hours, or even days, without sustaining either temporary or permanent injury. Short and easy trips into the cave have been known to effect a cure in chronic dysentery and diarrhea, when all other measures had failed.

In the case of those consumptives who spent several months in the cave, it is probable that what they may have gained from a stimulating atmosphere of uniform temperature, was fully counteracted by the absence of solar influence, and while they grew no better, they were probably no worse than they would have been had they never entered the cave. They grew rapidly worse, however, after leaving the cave, for want of the stimulus to which they had become accustomed, as they all succumbed so soon after returning into the less bracing atmosphere of the outer world. This is conclusive evidence that the absence of light favors the development of the tubercular diathesis.

Hopeful, as consumptives generally are, the prospect could not have been very pleasing at best. Just imagine a company of more than half a score of cadaverous looking beings, ill favored and thin as Pharaoh's lean kine, wandering hither and thither in darkness, or like specters in the dim light of their tapers, breaking now and then the almost painful stillness by a still more painful and sepulchral cough! Can such a gloomy prospect give rational promise of benefit to the patient?

AMERICANISM.

While we are trying to divest ourselves of the sad impressions left by the above picture, we will just step into this dreary old hospital ward of a cabin, and take a look at the inside. Every crack is made the receptacle of cards of business firms in every city in the Union. The floor is littered over with a legion of ghastly fragments, which, like the dead in Westminster Abbey, were obliged to yield their places, to make room for their successors. After a time these too must yield to the overwhelming torrent that follows. An agent of a well known sewing machine manufacturing establishment, obliterated at a single stroke a whole legion of the smaller fry, by stretching up a circular of mammoth proportions.

THE STAR CHAMBER.

A little farther along the avenue enlarges in every direction, assuming the dimensions of a vast elongated dome, reserving its characteristic entablature, architrave frieze and cornice complete, but on a proportionately grander scale. The concave ceiling is covered with a dark incrustation of iron and manganese, interspersed with shining crystals of gypsum or diaphanous selenite, giving the whole, in the dim light of our lamps, a striking resemblance to the starry sky. Though a miniature edition of the great starry universe, the illusion is wonderfully complete. Still we feel assured that this is not our native sky, for not a single familiar constellation appears above the horizon.

By consent of the party, the guide will here assign us the most eligible point from which to witness a rare exhibition of light and shade. He takes all our lamps and passes slowly away, but in such direction as to throw the combined force of the lights upon the high vaulted ceiling, thus bringing out beautiful constellations of a wonderful variety of magnitudes. When the proper angle is reached, the rays of light are reflected from an elongated, wing shaped crystal, producing a beautiful comet of great size and brilliancy. The likeness of the starry heavens is now so complete that we gaze in mute astonishment. The guide passes, by detour, behind the rocks, leaving us in the blackest of darkness, to enjoy our own reflections. Though we can hear the muffled throbbing of our hearts, as the vital fluid is forced through the system, the death like stillness soon becomes painful.

The minutes begin to drag their weary lengths, and we begin to feel as if the whole machinery of time were moving slower and slower, with a constantly retarding velocity. Just as it is about to stop, to our great relief, we catch a gleam of dim twilight, looming out from a frightful looking chasm on our left casting the shadow of a projecting rock upon the opposite wall. See its spectral visage, like the ghost of Hamlet, stealthily approaching us! Ah, it is gone! and all is again inky darkness, worse than before. What! was this an illusion? one of those incoherencies of mind which it is said invariably attack and bewilder persons lost in the cave! No, it was the light of the lamps shining through an aperture of the rocks, giving us a gleam of hope as they passed! Now we begin to attach a value to daylight which we never appreciated before.

The guide, with all his lights, has again appeared in full view at another aperture farther beyond, from which point he again entertains us with a beautiful stellar exhibition and dissolving views. Arranging a part of the lights so as to enable us to take in the dim dimensions of this vast dome, giving everything an uncertain, ghost like appearance, he again retires with the rest, and approaches us by a different route, so as to cast the shadow of a projecting rock upon the ceiling, simulating an approaching thunder cloud; receding again slowly, the cloud again vanishes, leaving the sky clear, calm and serene as in a mild summer evening.

We would fain linger here and gaze upon this miniature firmament, but the guide intrudes upon our delightful reverie, by announcing that it is time to take up the line of march.

FLOATING CLOUD ROOM.

This is a magnificent hall, a quarter of a mile long, corresponding with the Star Chamber in width and Night. The appearance of clouds is produced by the scaling off of the black gypsum, exposing the white surface of the sulphate of soda beneath. The effect of this illusion is wonderfully charming. The clouds seem to float from the Star Chamber over the Chief City.

PROCTOR'S ARCADE.

This wonderful tunnel is beautiful beyond description. It is one hundred feet wide, forty five feet high and three quarters of a mile long. The ceiling is smooth and the walls vertical, as if built by a mason. When this arcade is well lit up we feel that the descriptive powers of a Seneca could not do it justice.

KINNEY'S ARENA.

This hall, a hundred feet in diameter, contains a curiosity about which there has been much speculation. It is a stick of wood, several inches thick and three feet long, projecting out of the ceiling many feet above the floor. How this stick came there, or when, must long remain an unsolved problem.

WRIGHT'S ROTUNDA.

This large hall is beyond the S Bend. It is four hundred feet in its shortest diameter. The ceiling is about level, but the floor inclines so as to make it at one side about ten feet, and at the other forty five feet lower than the ceiling.

NICHOLAS' MONUMENT;

At the eastern part of Wright's Rotunda is a column four feet in diameter, extending from the floor to the ceiling. This is known by the above name in honor of Nicholas Branford, the oldest cave guide now living, and still to be found in the vicinity of the cave.

FOX AVENUE

is about five hundred yards in length, and communicates with the S Bend and the Rotunda. Some distance beyond Wright's Rotunda the main cave sends off several avenues. That leading to the left leads to the Black Chamber, so called because the walls and ceiling are covered with a crust of black gypsum. This hall is one hundred and fifty feet wide and twenty feet high. It is gloomy enough to suit the most despondent of rejected lovers.

Two avenues put off from this dark room. The former communicates with

FAIRY GROTTO.

This is a mile in length and contains a grand collection of stalagmites. The other communicates with Solitary Cave, at the entrance of which there is a small cascade.

THE CHIEF CITY

is situated in the main cave, beyond Rocky Pass. It is about two hundred feet in diameter and forty feet in height. The floor is covered with piles of rock that present the appearance of the ruins of an ancient city. Three miles beyond this point the cave terminates abruptly. In other words it is blocked up by rocks that tumbled down from the ceiling, as at the end of Dickson's Cave. The Cave does not end here. Who ever will dig through the barrier may be well paid for his trouble. No mortal eye has ever beheld what lies beyond these rocks. It is not known for a certainty that any rocks have fallen from the ceiling in any part of the cave since the first white man ventured into it beyond sight of the entrance.

BACK TRACK.

Again we retrace our steps through this grand panorama of wonders, pass the Deserted Village, turn around McPherson's Monument at the acute angle, take our bearings at the Giant's Coffin, and move toward the entrance. In about fifteen minutes, or perhaps less, we begin to recognize familiar objects. The old pump logs and leaching vats on the left, remind us that we are near the entrance of the

GOTHIC ARCADE.

Here we ascend the flight of steps fifteen feet high, on our left, enter an open passage leading off at right angles from the main cave. About forty yards beyond these wooden steps, in the left hand wall, is a little niche called the Seat of the Mummy. In this niche was found the body of a squaw, as dry as a herring. She was dressed in the skins of wild animals, and ornamented with trinkets usually worn by the aborigines. A few feet from the niche was found the body of an Indian child, similarly dressed. When discovered it was in the sitting posture, resting against the wall. How long these bodies had remained there is not known. When found they were in a state of perfect preservation. For a time they were exhibited through the country as curiosities, and finally deposited in Peale's Museum. We pass for the distance of a quarter of a mile, beneath a low, smooth rock, resembling the ceiling of a plastered room. Now we enter a considerable hall, that had originally a beautiful white coating of lime. This is called the

REGISTER'S ROOM

because the smooth rock over head is everywhere smoked over with names and rude characters or pictures of animals. These were placed here a long time ago. The proprietors very justly prohibit all such improprieties from being committed in these halls now. No candles are allowed to be taken into the cave. It is not long since that imploring appeals were made for candles, instead of lamps, but the applicants were rebuked by peremptory refusal.

This part of the cave, like the others, leads off in zigzag directions. It is stated by the guides that it even crosses the main cave on a different plane; though this is not certain, as no surveyor's compass is permitted to enter the cave, and for reasons we need not state.

GOTHIC CHAPEL.

We next reach a picturesque little dome full of wonders. The arch forming the roof is about thirty feet above the floor, supported by pillars of stalactitic formation so disposed as to make gothic arches of great beauty and regularity. How came these pillars here? Some of them are thirty feet high and from one foot to three feet and upward in diameter. More than half a dozen of these pillars are in this hall; but whence came they? It is worth the while to study them a little. Examine them closely and we find they are composed of carbonate of lime in crystals; or, in other words, crystallized limestone. We must come to the conclusion that they were carried here, set up in their places, and then cemented by water. At first this view seems unnatural and unreasonable, but most probably it is the only rational view after all. We have seen that the water, under certain circumstances, will dissolve the limestone and hold the fine particles in solution just as it will hold sugar or salt in solution.

We also have shown that the cave was formed by the rocks being actually dissolved by cold water, and the particles of lime being carried away in a state of solution. It is this water, impregnated with dissolved limestone, that constitutes what is known as hard water. If we take a strong solution of salt and allow the water to evaporate, the salt will remain in beautiful crystals. If this solution of salt be allowed to drop at great intervals from a string, acting upon the principle of a siphon, there will form a pendant mass of crystalline salt about the string; this is called a stalactite. Just beneath the string there will be formed a crystalline mass of salt growing upward. This is called a stalagmite. But this stalagmite will not be formed if the intervals between the drops are not sufficiently great to allow all the water of the first drop to evaporate and the salt to become crystallized before the occurrence of the next drop. If the drops fall a little too frequently they will melt away the stalagmite rather than occasion it to grow.

These immense masses of limestone, weighing more than ten tons, were formed just in the way above described. The lime, held in solution by the water, becomes crystallized under favorable circumstances. The pure water is taken up in an invisible form by the air, and whatever salt was contained in the water remains behind in a crystallized form; hence, rain water is always soft or free from salt. In the still air of the cave, perhaps one drop in twenty four hours is quite as fast as the water could evaporate and form crystals of sufficient hardness to withstand the shock of the next drop, without being reduced to the liquid state again. The amount of lime in a single drop of water must be small when crystallized. It is more probable that one drop per week is quite as frequent as is consistent with the growth of a stalagmite, or even one drop per month might not be too slow to favor the most rapid growth. It will be borne in mind, if the drops occur only a little too frequently to favor the growth of a pillar, they will occasion its destruction. It is said that the growth of one of these pillars, in the space of thirty years, is about equal to the thickness of a sheet of writing paper. It requires about thirty sheets of paper to be equal to one eighth of an inch in thickness, or two hundred and, say, fifty sheets to be an inch thick. Now, if it requires thirty years to grow the thickness of one sheet of paper, it will require two hundred and fifty times thirty years (seven thousand five hundred years) for the growth of one inch of the pillars. The growth of a single foot would require, at this rate, ninety thousand years. As some of these pillars are thirty feet high, it is fair to suppose that they grew half the distance from above and half from below in the same time. At this rate, it would have required nine hundred and forty thousand years for the completion of one of these shady old props.

If we bear in mind that these caverns have all been formed by the slow process of solution at low temperature, and that this process had ceased in this immense hall ere the foundation of these pillars was laid, we must come to the conclusion that mother earth is entitled to reverence for age beyond what is usually paid to her. The Florida reefs afford another instance of proof of the great antiquity of this terrestrial ball.

PERFORATE STALAGMITES.

Sometimes little pendant masses of limestone, resembling icicles in shape, are found with a hole passing through them, from end to end, straight and smooth as if bored with a gimlet. These perforations are probably formed as follows: A drop of water, holding in solution a large quantity of lime, remains suspended from the same point, and as its supply is just equal to the evaporation, it never falls. By and by, a little ring is formed around the base of the drop, and since there is no evaporation inside of the ring, it can not be filled up by crystallization. A series of these rings is formed on the same principle, and the tube is constantly full of water, receiving its supply from that, that runs down the outside to the opening at the apex, where it is forced in and held by atmospheric pressure. As long as evaporation continues the tube is prolonged downward.

THE BRIDAL CHAMBER.

To young folks, it is an interesting little story that suggested the name of this hall. It runs thus: A romantic young lady was here joined to the partner of her choice, notwithstanding the promise she made to her maternal parent, on the latter's death bed, that she would not marry any man on the face of the earth, and she didn't.

OLD ARM CHAIR.

This is a large stalagmite, resembling an arm chair. It is considered one of the curiosities of the hall. There are, also, a great many others, worthy a passing notice, such as the Elephant's Head, Vulcan's Shop, Napoleon's Breastworks, the Pillars of Hercules, and Gatewood's Dining Table. These are all stalagmitic formations or incomplete pillars, their stalactites hanging over them. Many of these pendant masses were recently joined to their stalagmites, but not very firmly cemented; others may yet be joined together ages hence. It is pretty evident that the pillars were formed in the manner described, as they are all thinner at the middle than at the extremities.

LOVER'S LEAP.

Just a little beyond these interesting pillars, the floor is let down about fifty feet, and a long point of rock projects out over the Devil's Cooling Tub, and innumerable and unknown caverns beneath. We will now retrace our steps, and in half an hour find ourselves descending the wooden stairway into the main cave.

Our day's work is almost complete. We have just seen the three great points of interest that are universally acknowledged the grandest caverns of the kind in the world. These are the Star Chamber, Gorin's Dome, and Stalagmite Hall. Whatever of grandeur or beauty they may possess entitling them to the distinction of masterpieces, they still lack the ornamental embellishments of those enchanting palaces beyond the rivers. But of these we will have more to say on the morrow! We have now been in the cave about four hours, and as we enter the narrows, the outward current of air begins to play a hazardous game with our enfeebled lamps, but as we are in sight of day light, we allow them to be extinguished.

END OF THE SHORT JOURNEY.

As we approach the little cascade, its peculiar humming sound strikes the ear with a clearness not noticed before, and the sky seems to wear a brighter hue. We should linger some minutes at the floor of the cavern, in order to allow the eye and the lungs to become accustomed to the new conditions, before ascending into the bright sunlight and warmer air. After ascending the steps at the mouth of the cave, it is well for the tourist, especially if he have weak lungs, to rest at least ten minutes within a few yards of the break where the cool air of the cave is quite distinctly felt, thus becoming gradually accustomed to the warmer air of the outer world. If this precaution be neglected, sudden and severe prostration may follow.

The line between the cool air of the cave, and the warm air of the outer world is so accurately maintained that we may stand in the one and thrust the hand into the other, and feel the difference of temperature quite distinctly. The shadows of the trees and their leaves are noticeably dark, forming a strange contrast with the bright sunshine. The harsh jargon of myriads of crickets and other insects, and the chirping of birds are soon lost upon the senses. The pupil of the eye, on coming out of the cave, is wonderfully enlarged, hence the strange appearances of the light and the shadow. If the eye be a strong one, it will soon correct itself, and everything will wear its natural appearance again.

VEGETABLE ODORS.

Persons who have been long at sea, can smell the land when they approach very close to it. Coming out of the cave the grasses, flowers, and leaves seem to exhale odors of wonderful sweetness. In some persons the sense of smell has become so acute by remaining in the cave a few hours, that the romance of a pure country air is forever lost to them. Odors that were sweet to them before, have become distasteful, and in some cases even nauseous. This is very easily understood and needs no explanation. Ascending the hill to the hotel, we feel languid and oppressed with the closeness of the air, but this soon passes off, and we are all the better for being one day's journey the wiser.

OWNERSHIP.

The cave, with seventeen hundred acres of land, is owned by the Croghan heirs. At the death of the original owner, Judge Underwood, of Bowling Green, became their trustee. The judge being well advanced in years, is about taking the necessary steps to obtain a decree of court to enable him to sell the property. Should it fall into the hands of enterprising capitalists, it will probably become one of the most desirable, convenient, and interesting summer resorts on the continent.

The soil in this part of Kentucky is unsurpassed in productiveness. The peach and the grape grow to great perfection, and rarely fail. Mr. A. L. Mallory showed me a fine peach tree in his garden at Cave City, which I found on measuring to be six feet high and five inches around the tree at the ground. This tree sprung from the seed less than one year previous, and was consequently the result of a single summer's growth. Apples, too, and small fruit may be cultivated with profit. The seasons are mild compared with Southern Ohio. Spring opens about two weeks earlier in the regions of the cave than at Cincinnati. On the first of April the general appearances of Spring around Cave City were as strongly marked as they were at Cincinnati twenty days later.

NATURAL BEAUTIES.

This quiet retreat combines nearly all the advantages of a first class summer resort. There is a bountiful supply of pure, sweet water, fresh air, and sunshine or shade, as we prefer. We are free from the noise, dust and bustle of crowded thoroughfares, surrounded by extensive shady forests, high hills, and deep, cool valleys. The wild deer and wild turkey roam here at will, even at this day.

It is just the place for a pleasure seeker or invalid. Our every want is amply provided for by the obliging host. We are free to spend the evening as we choose, without conforming to the conventionalities of Long Branch or Saratoga. If we dance we will here enjoy it. In the absence of nearly every exciting cause, Nature claims her right, and we retire early, because we wish to. After the preliminary preparations, we bid the world good night, and Morpheus takes us gently into his embrace, and soars away to the land of pleasant dreams. Early next morning we are awakened from sweet slumber by the music of birds and the fragrance of roses stealing through the open window, and pervading our apartment.

The highest degree of bodily health and mental vigor is generally found in connection with that condition of the system which accepts no more than the least amount of recuperative agent, capable of satisfying its real wants. To have slept just enough is a great luxury, while too little or too much sleep is equally injurious. We dismiss the god of dreams at once, and exclaim with Sancho Panza, "blessed be the man who first invented sleep," and add, by way of variety, the whole feathered tribe, and the floral dictionary. If they do occasionally cut short our morning nap, they also bring us closer to Nature's self. Our loss is doubly restored in tasting the fresh, pure air, before it is tainted by the breath of any lazy man, and in witnessing the glory of an original sunrise.

Breakfast over, we prepare for the long journey.

THE GUIDES,

being amply freighted with oil, matches, lights and refreshments, provide us again with newly trimmed lamps, and we take up the line of march out through the garden and down the hill as before. The same sights greet us on every hand, and emotions similar to those of the previous day are awakened by the odors of fresh green leaves and the songs of the merry birds.

The morning is too fresh and fair to be exchanged so soon for the darkness of night, but the guide leads on and we must follow.

We will not go farther without saying a few words about the colored guides, Mat and Nick. Mat is a fair average specimen of the Anglo African race, improved by a considerable degree of culture, acquired by Contact with scholars, professors of every science, and especially of geology and mineralogy. During the thirty years in which Mat has acted in the capacity of cave guide, he has collected a vocabulary of scientific terms that would do credit to a man of letters. Sometimes he handles these musical phrases with as much skill and appropriateness in hewing down the barriers of ignorance as a mechanic who understands the use of his tools might display in swinging a broad axe.

Upon some occasions he is really amusing, as well as somewhat instructive. No one should fail to secure Mat's services as guide. He is slim, tall and wiry, and fifty years of age. He was for many years the associate of Stephen, the world renowned cave guide. Nick has also grown old in the service, and is still a faithful and efficient guide.

Mr. Samuel Meredith, a native of the county, has long been familiar with the cave, and is acting as guide at present.

During the visiting season quite a number of efficient guides will be employed, many of them making the trip every day. By this time we reach the floor of the cavern, and pass the cascade without fear.

DISCOVERIES.

Previous to the year 1838 the only part of the cave familiar to the explorer was that from the entrance to the Bottomless Pit. In this year some bold adventurer passed the hitherto limit of subterranean knowledge, and was rewarded by the discovery of greater wonders than had yet been seen; but the dangers incurred in reaching them increased in like proportion. So great and daring a feat, and yet so successful an enterprise, could not long go begging for a rival. In 1840 the rivers were navigated and crossed for the first time by living human beings.

A SECOND VIEW.

For about two miles we follow the path of yesterday. We feel a peculiar satisfaction in viewing a second time some of the curious sights we saw on the previous day. A little reflection, after a sound sleep, seems to disrobe many curious things of at least a part of their enchanted garments. We pass the Narrows, the first vats, the pump logs, and enter the Rotunda. So far we have the same impressions we had from the first, nor do we fear that such will not be the case with all the rest; but judgment sometimes runs against the impenetrable rock of reason, and we are compelled to change base. To illustrate: We may entertain a favorite opinion of this or that theory, and steer clear of all objections many times and yet fail in the next attempt. This fact was exemplified by one of the cave guides, who had made the trip successfully ninety nine times, and yet knocked himself down on the hundreth by running his head against a rock projecting from a low ceiling.

We will not stop to make new observations, but pass on to the angle of the main cave, one mile from the entrance. Having reached this point, we turn around the Giant's Coffin to the right, leave the main cave by creeping down toward the Deserted Chambers, and by some strange fatality, bump the sorest part of our head against the identical peak that reminded us so forcibly on the previous day of the old lady's instructions to Thales, viz: "Keep the head from among the stars while the feet are on the earth." If Hannemann, the father of Homoeopathy, had converted himself into a troglodyte, and passed down this way several, times, receiving each time a greeting similar to ours, what an innumerable quantity of bitter pills humanity would have been obliged to feed upon; for he would not have subscribed to that famous motto of his adherents, viz: "*Similia similibus curantur,*" (likes are cured by likes). Reassured that a repetition of the cause will not remove the effect, we pass on through the Deserted Chambers, by the Bottomless Pit, over the Bridge of Sighs, along the Arched Way, leaving the Winding Labyrinth and Gorin's Dome to the right, and enter Revelers' Hall. In this wild, irregular, triangular cavern, we might easily mistake the way, and traverse Pensacola Avenue, pass beneath Snowball Arch and enter Wild Hall, but for the sagacity of the guide, who leads us gently as lambs right into the Scotchman's Trap. We take another look at this arrested deadfall, and wonder at the slender thread upon which many of the chief pleasures of life are suspended.

Satisfied that there is no danger of the Trap being sprung while we are in it, we pass down the steps, begin to do penance by stooping, for we have entered the Valley of Humility. This terminates by entering, at almost right angles, a higher passage, the part of which leading to the right being known as Bunyan's Way, and that to the left Buchanan's Way. We follow that bearing the name of our late chief magistrate. Just forty yards from where we entered this grotto we turn off to the right into Fat Man's Misery. If we are a little stout we must keep the channel, lest in striving to escape Scylla we wreck upon Charybdis.

After threading this sinuous passage its whole length, we are thrust into Great Relief, with a feeling somewhat akin to that experienced by the successful applicant for a post office, after having worked his way through the crowd and gained admittance to the audience room. Here we take a breathing spell, and experience great relief in another sense. Passing on now to the right, we soon enter River Hall, and the Bacon Chamber. A little further forward is the Dead Sea, whose dreary depths are as calm as if oppressed with the gloom that constantly hovers over it. The river Styx probably ran through this quiet pool before the channel was cut to its lower level.

Time has changed the character of these pools and streams, but most probably the changes will not be so apparent in the future; as the streams in the cave, being connected with Green river, will never find a lower level than that of the aforesaid stream. Off to the right is a high, open grotto, that may be followed for miles in its zigzag directions, crossings and windings. At one place there is a grotto of great size, cut right across another, and considerably deeper, just as a furrow in a cornfield is cut across and deeper than another. This is a strange freak, and the only instance of the kind we have seen. In the deeper furrow, a troop of cavalry might ride off in either direction. This passage is not shown to visitors. An open passage leading forward from River Hall to the River Styx will give us access to Charon's Boat. This fabled craft has become rather old and frail, and the old gentleman has very wisely hauled it off. Parties may still take a trip on the River Styx in a boat of modern build and ownership. Viewed from the bridge that spans this river, the effect of the spectral scene, reflected from the dark waters of the fabled stream, is strange indeed and impressive beyond comparison. As the party approaches the shore, a Bengal or Calcium light heightens the beauty of the scene.

PASSAGE OVER THE STYX.

In mythological times, Mr. Charon seemed to monopolize the ferry business, and so unpopular did he become that no one patronized his ferry, except under protest, and I believe to this day there may be found plenty of tourists who would rather swim the Styx than step into the old curmudgeon's boat. Happily we are not left with this alternative, for we have a natural bridge spanning the Styx in the cave, and over this parties usually pass, without paying any toll. We take the road leading to the left in River Hall, and pass over the bridge. Part of the time our path lies close to the right of the grotto, and then we cross over a narrow bridge to the left. Here there should be put up a firm iron railing to provide against every possibility of danger, though no one has ever been injured beyond a fright. We stoop in order to go safe, and creep along about thirty yards when we cross over to the right and descend a steep stairway that brings us down nearly on a level with the Dead Sea.

Passing on a little further we begin to go up grade and enter a passage somewhat lower and narrower. A few rods further and the ceiling springs ninety feet over our heads as we descend the slippery bank and stand on the shore of the Lethe.

THE EMBARKATION.

Thick, dreary darkness is before and behind us; the dark gray rocks on each side and over us, and at our feet lies a pool of water, whose surface never rippled by the breeze, looks calm as the face of an honest man in deep meditation. Moored by a stake driven into the sand, is a little boat, narrow, short, and shallow. This is the craft in which we intrust our most sacred treasure for a voyage over the Lethe. Shall we risk the voyage? Why not? The waters are calm, and there is not motion enough in the atmosphere to turn a feather. In these dreary realms no tempests hover over the waters. Aeolus is locked up in his rocky cave, and can no more invade these dominions than can the rays of the sun. Why not risk the voyage in this craft? Have we not entered upon the voyage of life in a bark infinitely more frail, without chart, compass, or experience, and the ship not so much as provisioned for a single day, and every foot of the way beset with lurking dangers, tempests, strife, bickerings, jealousies, irreverence, dishonesty, and, above all, unkindness to ourselves? Alas, poor humanity is always straining at a gnat.

With some such feelings we step into the boat, freighting her down within a few inches of her bearings. The guide now plies the paddle steadily, and skillfully, for we are entered upon a new field of explorations. The lights in the prow cleave the thick, murky darkness that closes around behind us, as the waters close around a fish. As we strain the vision to look out into the dismal realm, like a wrecked mariner looking out for a friendly sail, or for land, an impressive feeling creeps over us and we meditate upon the singular relation we sustain toward the world and ourselves. We are afloat upon a river that the sun never shone on, far beneath the surface of the earth, and in the enjoyment of life, health and reason, and yet cut off from all communication with the outer world, except through the medium of the little boat and our lamps.

THE GREAT WALK.

Having gone fifteen or twenty yards, the perpendicular wall on our left gives way to a low sandy beach. On this we disembark and fasten the boat carefully to a stake placed here for the double purpose of anchor and landmark. Our path now leads over undulating sand deposits, many of them very large. In some of these the sand is very fine and white in others it is coarse and red. These heavy deposits seem to indicate a powerful current when the rivers are high. All along this part of the grotto the ceiling is high over our heads. It seems probable that the rivers began to flow between strata of limestone rock ages ago, cutting away the lower rock and leaving the upper which now forms the roof at various distances above the floor. The same characteristics obtain here as elsewhere. In ten minutes from the time we leave the boat at the Lethe, we come to the

ECHO RIVER.

This wonderful stream is the most interesting object in this part of the cave. Its water has a temperature of fifty four degrees, and at some seasons of the year is so transparent that the bottom may be distinctly seen at a depth of fifty feet. This stream is navigable for nearly a mile. It is about one hundred and fifty feet wide and contains water enough to float the largest East river or Mississippi steamer.

A TRIP ON THE ECHO RIVER.

We step into the square little boat and arrange ourselves upon the seats, consisting of boards laid across the gunwales. The vessel being properly trimmed we push off from her moorage and float gently with the current of the stream. Now the little craft is at the mercy of deep, silent water. On each side is a perpendicular wall a hundred feet high, vaulted over head by solid masonry. On our right, more than half way from the water up the wall, a huge cornice projects ten or twelve feet out over the stream. This cornice is concave beneath and ornamented with pendants resembling fretwork. On our left the wall gives way to a great deposit of fine sand, sloping gradually upward from the water's edge till it joins in places the descending roof at various distances from the river. The same outline is preserved, so far as we are enabled to penetrate the darkness in either end of this vast corridor. Between these sand deposits on our left are several avenues leading off from the river. These communicate with our path further along. One of them passing over a high mountain of sand and rocks promiscuously jumbled together is called Purgatory. This path is followed by the explorer when the river is too high to pass beneath the arches.

The peculiar feature of this region is the powerful echo that responds to every sound. One almost fancies that the susceptible nymph, whose excessive love for Narcissus caused her to shrivel up and pine away till nothing remained but her voice, had taken up her abode among the beetling rocks overhanging this river.

Whatever the renowned and valiant Don Quixote may have experienced during his enchantment in the famous cave of Montesinos, we doubt much whether it surpassed in sweetness the angelic chorus into which this nymph's voice still converts the simplest sounds for the entertainment of tourists. On our first visit to this place we would have been quite spellbound but for the chagrin we felt at having left our flute behind.

We pass along the dark shore, avoiding all purgatorial experience, when presently the river seems to have come abruptly to an end; but not so, for on our right there are a few feet of space between the rocky arch spanning the river and the surface of the water. Through this we are doomed to pass. We reverently lower our heads and guide the motions of the boat by pushing with our hands against the rocky firmament arching over us. After going a few lengths of the boat we are again unable to reach the roof, for it springs up as suddenly as it was lowered. We can stand up at full length and only occasionally touch the rocks beneath which we are carried with the accelerating velocity of the stream. After going half a mile or more our ears are greeted by the unwelcome sound of falling water. As we rush on, louder come the doleful cluckings of the waves that roll into the little coves on both sides of the stream. Faster and faster sweeps our boat into this Hellgate; louder and fiercer come the sounds from the cascade, till they begin to excite alarm. The guide keeps on as if he neither saw nor heard aught, and but for his coolness we might become desperate under the suspense. Suddenly he turns the boat into a little rocky inlet on the left side of the stream, where we disembark and fasten the craft securely to her moorage.

Should our boat by any mishap get away from us here, or our lamps be extinguished, or our oil be consumed, without any means of supplying the want, we could no more find our way out than fallen man could find his way to a better world without the lights of Christian faith and revelation. Though fogs sometimes arise here, no misfortune has yet befallen any tourist. The guides understand their business too well to allow any such thing to happen. They never attempt the long journey without having made ample provision for every contingent emergency. Should a party remain in the cave beyond the proper time, a guide is sent in search of them. But such a necessity has not arisen in the space of many years.

CASCADE HALL.

After climbing over a rough ledge of irregular rocks we finally reach dryer and safer footing. We pass around a sharp pillar of dark gray limestone and are ushered into a romantic cavern. Judge of our surprise at discovering the cause of our consternation, a small stream of water, not much thicker than a gentleman's cane, issuing from an orifice in the ceiling and dashing upon the rocks some thirty or forty feet below, and finally disappearing through a funnel shaped hole in the floor.

This is the wildest and most romantic cavern beyond the Styx. It has an area of from a quarter to half an acre, irregular and picturesque on every side. The floor is almost impassable with rocks and rubbish that have fallen at intervals from the ceiling. "It is a fit habitation for the gnomes, who until recently held unenvied and undisputed possession.

SILLIMAN'S AVENUE,

a passage, seven or eight yards wide, from fifteen to twenty yards high, and a mile and a half long, begins at Cascade Hall. The floor is very irregular, owing to portions of the ceiling having tumbled down from time to time, perhaps centuries ago. Either wall has a well defined cornice of shelving rock, near the roof of the cavern, which gives it a heavy architectural finish. On each side of this grotto little avenues put off, wind through the rocks, and finally return to the main cave again.

DRIPPING SPRING.

This is a pool of water receiving its supply from the ceiling. A slippery place close by is known as the Infernal Regions. The Sea Serpent is a tortuous crevice cut by the action of running water into the ceiling overhead. Beyond the Hill of Fatigue is the stern of an immense ship, with her rudder hard-a-port. Next is the Rabbit, a large stone resembling in shape the animal whose name it bears.

OLE BULL'S CONCERT ROOM

is on the left of Silliman's Avenue. It is thirty feet wide, forty feet long, and twenty feet high. When the inimitable violinist made his first tour through the United States, he visited the cave and performed in the hall that still bears his name. He probably gave this hall the preference, because the proportions of its length, width, and height are nearest to those of a hall built in conformity with the well known laws of acoustics.

RHODA'S ARCADE.

This is a charming hall putting off from Silliman's Avenue, about one mile from Cascade Hall. It is five hundred yards in length and from five to ten feet in height. The walls and ceiling are covered with crystals of gypsum, and the floor is strewn with brilliant fragments that have been detached from their place of formation.

LUCY'S DOME.

This is the highest dome yet discovered in the cave. It is more than three hundred feet from the floor to the top, while its lateral diameter is only about sixty feet. Its mural embellishments are one continued series of curtains from the top to the bottom. It may be reached by passing through Rhoda's Arcade.

About one mile and a half from Cascade Hall, a steep bluff divides the avenue into two separate grottoes; that on the left still bears the name of Silliman. We may follow it over a mile further without coming to any definite result. Tons of fibrous gypsum, white as snow, and fragments of alabaster of various colors are found in great abundance among the shelving rocks. Leading to the right is the

PASS OF EL GHOR.

For the distance of a mile and a half, El Ghor is a distorted labyrinth of beautiful and surprising sights; now narrow and lofty, now flattened out between horizontal strata of limestone, whose broken edges assume the most remarkable forms.

Here is a little vestibule with moldings and friezes of the gothic style of architecture; beyond is a Cretan labyrinth of the most singular and uncouth proportions, terminating in a series of ramifications leading to several tiers of avenues.

It were too tedious to describe every particular object in detail, therefore we pass over to the points of greatest interest and beauty. Among these are the Hanging Rock; the Fly Chamber, in which a profusion of specks of black gypsum resemble a legion of flies upon the walls and ceiling. Table Rock is twenty feet long and projects ten feet from the side of the hall. It is only two feet thick.

The Crown is about six feet in diameter and ten feet from the floor, on the right side of the hall.

Boone's Avenue, leading off to the left, has been explored about a mile. It contains nothing of interest. Corinna's Dome, forty feet high and nine feet wide, is just over the avenue. The Black Hole of Calcutta is on the left side of the avenue. Stella's Dome is two hundred and fifty feet in bight. It is reached by an avenue in the left wall of the Pass of El Ghor.

A series of pending rocks in this part of the hall is called the Chimes, because they emit a sound when struck by the hand.

HEBE'S SPRING

is four feet in diameter and a foot and a half deep. The water contains sulphur. Half a mile beyond this spring the pass communicates with Mystic River, the extent of which is unknown.

MARTHA'S VINEYARD.

We will next climb up this ladder about twenty feet, and creep through a hole in the ceiling large enough to admit one at a time without any inconvenience. Now the passage leads gradually upward, as if it had been dug out by rabbits or marmots for their special habitation. Here a large vine springs up the wall from the base, reaching nearly to the top, where it supports a dense mass of foliage and clusters of grapes of wonderful size, their rich tints of blue and violet shining through the water that trickles over them.

The plump, shining fruit, forever ripe and forever unplucked, clusters so thickly together as to hide the leaves of that subterranean vintage.

WASHINGTON HALL.

A little way beyond the vineyard, the grotto turns to the left and becomes more level on a higher plane. Here we enter Washington Hall, a beautiful Circular dome, over one hundred feet in diameter. The subdued color that lingers about the periphera of this hall, in spite of our lights, gives it an air of wonderful antiquity, almost too sacred to be devoted to no holier purpose than that of a dining hall, and yet this is the almost infallible practice of visitors. Cans of oil are kept in this room, from which the lamps are replenished while the tourists are regaling the inner man. Marion's Avenue begins at Washington Hall and leads to Paradise, Zoe's Grotto, and Portia's Parterre.

ELINDO AVENUE.

Here begins this splendid avenue, about twenty yards wide, six or eight yards high, and two miles long. This is a charming hall, white as snow, dotted with sparkling crystals of gypsum. The diamond grotto and gardens of sparry efflorescence in the gleaming vaults of this magic hall, almost bewilder the senses.

THE SNOWBALL ROOM.

This is a beautiful hall, reminding one of the sports of yore. The walls and ceiling are mottled with spots resembling those made by throwing a snowball against a solid wall. We almost fear to examine the little pyramids, lest they should soften by the heat of our lamps and drop off. Surely the gnomes must have been a rollicking set of roysters when they were yet younger. Latterly they have grown very sedate and shy, for I am informed that not one has ever allowed itself to be either seen or heard. Next is

CLEVELAND'S CABINET

of Crystals, whose brilliant scintillations and wonderful richness captivate us as completely as did the diamond eyes of Juggernaut the Hindoo. This magnificent hall is a mile and three quarters long, sixty feet wide, and from ten to twenty feet high. Its variegated and brilliant display surpasses anything we have yet seen. The following are some of the finest parts of it: Mary's Bower, fifteen feet high and forty feet in length. The walls and ceiling are covered with white rosettes. The Cross, two crevices intersecting each other at right, angles, twined with flowers of gypsum. Mammary Ceiling, nipple shaped projections. The Last Rose of Summer, about eight inches in diameter, pure alabaster and white as snow. Bacchus' Glory, a little alcove lined with nodules of gypsum, resembling grapes. St. Cecilia's Grotto, Diamond Grotto, and Charlotte's Grotto, are all too brilliant and enchanting to be described. All these formations bear a wonderful similarity to branches of vines, leaves and flowers in the organic world.

THE FLOWER GARDEN.

Here is a conservatory for inorganics, stocked with the indigenous flora as well as with exotics. White roses and sunflowers, daises and lilies, the convolvulous, and the feathery chalices of the cactus hang in bewildering profusion from the crevices. The night blooming cereus flourishes here without fear of the advancing morn blasting its fair fame. The scarlet of shame never blushes the tulip, for it is here a virgin pure and white as snow. Long, snowy, pendant wreaths of cactus flowers trail their fringes upon the floor. The passion flower, the iris, and bunches of celery, that might tempt the appetite of a fairy, flourish and sparkle on every side. Rock blossoms, fair as alabaster, mingled with those of every shade of color, vegetate in great profusion upon the walls and along the crevices in the ceiling.

ROCKY MOUNTAINS.

From the farther end of Elindo Avenue may be seen a huge pile of rocks, heaped one above the other to the height of a hundred feet. As we gaze upon their dim outlines, we may easily imagine them to be miles away, and their height to be equal to the height of the Andes. Far above these, like a dark cloud in the night, may be seen the ceiling whence these rocks tumbled in promiscuous ruin down.

CLEOPATRA'S NEEDLE.

On the very top of this mountain there is a pretty little stalagmite, two feet high, and six inches in diameter. This shows that these rocks must have tumbled down ages ago.

DISMAL HOLLOW.

Beyond this mountain is an unexplored valley, named by Stephen (its discoverer) Dismal Hollow. What beauties or wonders, fraught with abundant dangers, lie beyond this point, must long remain subject to conjecture.

At the mountain the cave divides into three branches. That leading to the right terminates in Sandstone Dome. The sandstone found in this dome may be taken as evidence that the top of it reaches near to the surface of the earth. Another evidence which seems to corroborate this, is, the existence of numerous lizards, crickets, and rats in this part of the cave.

CROGHAN'S HALL.

Some distance along the ridge of the mountain, is the entrance of a cavern opening to the left; this leads to Croghan's Hall, in one side of which the floor has dropped down into a deep pit, in which darkness reigns supreme.

FRANKLIN AVENUE

is the central path and leads from Dismal Hollow to Serena's Arbor, a distance of a quarter of a mile.

SERENA'S ARBOR

is a beautiful cavern, twenty feet in diameter and forty feet high. The walls and ceiling are highly ornamented with crystals and semi-transparent stalactites. These being sonorous, give out a musical tone when struck by the hand. This is called the music of the cave. In the awful stillness of this dreary region this music produces a singular effect.

THE MAELSTROM.

This is a pit thirty feet in diameter and of unknown and unexplored depth. The openings of avenues are visible at considerable distance down the pit, but these have never been and probably never will be explored. In the summer of 1859, William Courtland Prentice attempted to explore this region by descending into the Maelstrom. He was let down in a basket attached to a rope arranged with pulleys. The working of the apparatus was then entrusted to the management of some young friends of the bold adventurer.

Several accounts of this perilous descent have been published. It has been made the thread of a spirited narrative poem, by George Lansing Taylor, from which we take the following extract. The poem is entitled

IN THE MAELSTROM.

"Down! Down! Down!
Into the darkness dismal; Alone, alone, alone
Into the gulf abysmal,
On a single strand of rope,
Strong in purpose and in hope,
Lighted by one glimmering lamp,
Half extinguished by the damp,

Swinging o'er the pit of gloom,

Into the awful stillness,

And the sepulchral chillness,

Lower him into the Maelstrom's deeps,

Where Nature, her locked up

Mysteries keeps;

Lower him carefully,

Lower him prayerfully,

Lower and lower and lower,

Where mortal never hath been before,

Till he shall tell us, till he shall show,

The truth of the tales of the long ago,

And find, by the light that the lamp shall throw,

If this be the entrance to hell or no."

In descending, the adventurer encounters a waterfall, or cascade, which is thus beautifully described:

"But behold from rocky wall,

Circling round the shaft below,

Spouts a crystal waterfall;

All its coarseness,

And its hoarseness,

When he sees how fair their service is,

Vanish, till, by aid of vision,

Sounds infernal grow elysian.

Now he swings anear the side

Of this weird and wond'rous tide,

Where its limpid billows slide,

And its sheets, descending glide,

Veiled in whiteness like a bride;

Glistening where his lamp is beaming,

Sparkling, flashing, glittering, gleaming,

Like a shower of diamonds streaming

From the lap of Nature dreaming;

Streaming downward, passing quickly,

Sprinkling now upon him thickly,

From the fissure far above him,

As if all the Naiads love him,

With so rich a love and tender,

That they shower baptismal splendor;

Floods of jewels for his visit

Is't a flood of gems? or is it

That their kisses almost drown him?"

Enchanted by the beauty of these fearful depths, the young hero

still demands to be lowered

"Into the dark profound,

A deep that ne'er did plumbet sound;

Still he descends,

And anxiously bends,

Gazing down in darkness that never ends

Whose dimness, And grimness,

And darkness, And starkness,

And deepness, And steepness,

And deadness, And dreadness,

More frightful are made by his lamp's sickly redness;

Till checked by sudden shock,

He stands on solid rock,

Ninety and a hundred feet

From the friends who hold that cable;

Will they lift it, are they able,

Face to face, once more to greet?

He enters a hall,

A huge niche in the wall,

Where echoes unnumbered respond to his call,

From a roof that impends

Where a gallery extends,

Till, bounded by distance, in darkness it ends.

Now along its spacious flooring,

Eager, pleased, he roams exploring;

O'er obstructions, through wide chambers,

Onward still he wends and clambers,

Stalagmitic cones and masses

Glitter every where he passes;

Glitter through the gloom like glasses;

Shapes of beauty forming slowly,

Arches, shrines, and altars holy;

Groups of columns polyhedral,

Like some rich antique cathedral;

Nature's grand and 'gloomy glory,

Fairer than the fanes of story.

Thus he wanders,

Roams and ponders,

Through this gallery of wonders,

Till a rocky barrier rising

To an altitude surprising,

All across the chamber closes

And effectually opposes

All his efforts to get o'er it,

And he stands repulsed before it,

Yet he sees the cave extending

Onward till in distance blending

With the darkness, as if Nature

Were resolved to hold some feature

Hidden still from mortal creature."

These beautiful verses are thus vigorously closed, with, as might be expected, a promise of future greatness and glory, for the hero of this daring adventure:

"Down in that depth, where no other has trod,

Where writing was none save the writing of God,

Was graven a name

By that glimmering flame

That shall live on the record of daring and fame."

William Courtland Prentice, the hero of these verses, espoused the Southern cause, and was killed in an attack on the town of Augusta, Kentucky, in 1862.

END OF THE CAVE.

This is called the end of the cave. Though it is nine miles from the entrance, it is no more the end of the cave than is the last production of the sculptor or painter the end of art.

Some mathematical troglodyte has estimated the whole series of caves, grottoes, halls and rivers at one hundred miles in extent. Whether it is more or less, we are unable to determine.

We have now accompanied you to the distal end of the journey, and can not be so ungallant as to desert you here. We will guide you back leisurely to the entrance, pausing occasionally to glance briefly at what we had passed too rapidly.

EYELESS FISH.

These are of two varieties, and by no means plenty. In the one variety there are rudimentary eyes, and in the other no traces of such organs are seen. We saw a few of each variety, but found their size much less than we were led to suspect. It is said that specimens of these eyeless fish have been found six inches in length. We saw none over two and a half inches in length, and so transparent were they that every organ in the body could be easily seen with the naked eye especially if viewed in a strong light. For full scientific reports of these fish, see *American Journal of Science*, second series, volume seventeen, page 258, May, 1854, and volume forty five, page 94, July, 1843; and New York Journal of Medicine, volume five, page 84, 1845.

EYELESS CRABS.

These are also rather scarce, yet at low water some may be found by those who know just where to look for them. We saw a few that had been taken some time previous, and kept in an aquarium. These, too, seemed to be destitute of the power of hearing, but any motion imparted to the water seemed to create alarm. They probably have a high nervous sensibility, which is not inconsistent with their fair, soft, gelatinous appearance, in which they do not differ from the fish.

It is thought that both the eyeless fish and the eyeless crabs found in these subterranean rivers are viviparous, bringing forth their young in the living state, differing in this respect from other fish, which are oviparous, or egg producers (excepting the mammals, such as the whale and porpoise).

CRICKETS.

Of these we saw but one, and it was not more than half a mile from the entrance. It did not differ from the variety usually found in dark and damp cellars. Its eyes were small, yet it showed unmistakable signs of recognizing the light of the lamp when brought near it. As one long incarcerated in a dungeon, it had lost that brown, healthy color peculiar to animals that move occasionally in sunlight.

RATS.

These are exceedingly numerous or very restless, and more probably both, judging from the number of tracks every where seen between the entrance and the rivers, and we believe, too, beyond the rivers, as they most likely understand the kind of navigation adopted by the Teuton who came around the sea by land. We saw no rats, and there is no evidence that they are either blind or unable to hear, for they scamper away and hide at the approach of visitors.

MAMMOTH DOME.

Spark's Avenue extends from River Hall to Mammoth Dome, a distance of three quarters of a mile. This is a magnificent dome, two hundred feet in diameter, and two hundred and fifty feet high. Standing beneath its mighty arch as the guides light it up, we are impressed with the awful grandeur of the place.

BANDIT'S HALL.

This hall is about sixty feet long and forty feet wide. The floor is covered with large rocks that have become detached from the ceiling so long ago that no trace of their fracture remains above to show where they came from. To the right of this hall is an unexplored avenue of unknown extent. It is called Brigg's Avenue. Newman's Spine is a curious crevice in the ceiling, resembling in outline the vertebral column of some huge animal.

SYLVAN AVENUE.

This is about three hundred yards long, extending from Spark's Avenue to Clarissa's Dome. This avenue is remarkable for the great number of ferruginous billets of limestone, varying from five to fifteen inches in diameter. They look like petrified logs of wood, some of them having half the bark stripped off.

We have now been in the cave about eight hours, and it is time to retrace our steps to the entrance. Though we have traveled by land and water about twenty miles we feel as fresh and vigorous as if we had walked only a mile for exercise. Everything wears the same strange appearance as on the previous day.

We linger about the entrance of the cave in compliance with the caution and reach the hotel in half an hour more conscious of hunger than of fatigue.

Thanking you for the patience and forbearance with which you have so kindly followed us, we will take our leave of you by introducing the following beautiful poem:

MAMMOTH CAVE.
BY GEORGE D. PRENTICE.

All day, as day is reckoned on the earth,

I've wandered in these dim and awful aisles,

Shut from the blue and breezy dome of heaven,

While thoughts, wild, drear, and shadowy, have swept

Across my awestruck soul, like specters o'er

The wizard's magic glass, or thunder clouds

O'er the blue waters of the deep. And now

I'll sit me down upon that broken rock

To muse upon the strange and solemn things

Of this mysterious realm.

All day my steps

Have been amid the beautiful, the wild,

The gloomy, the terrific. Crystal founts

Almost invisible in their serene

And pure transparency, high pillar'd domes

With stars and flowers all fretted like the halls

Of Oriental monarchs, rivers dark

And drear and voiceless as oblivion's stream

That flows through Death's dim vale of silence, gulfs

All fathomless, down which the loosened rock

Plunges until its far off echoes come

Fainter and fainter, like the dying roll

Of thunders in the distance, Stygian pools

Whose agitated waters give back a sound

Hollow and dismal, like the sullen roar

In the volcano's depths, these, these have left

Their spell upon me, and their memories

Have passed into my spirit, and are now

Blent with my being till they seem a part

Of my own immortality.

God's hand,

At the creation, hollowed out this vast

Domain of darkness, where no herb nor flower

E'er sprang amid the sands, nor dews, nor rains,

Nor blessed sunbeams fell with freshening power;

Nor gentle breeze its Eden message told

Amid the dreadful gloom. Six thousand years

Swept o'er the earth ere human footsteps marked

This subterranean desert. Centuries

Like shadows came and passed, and not a sound

Was in this realm, save when at intervals,

In the long lapse of ages, some huge mass

Of overhanging rock fell thundering down,

Its echoes sounding through the corridors

A moment, and then dying in a hush

Of silence, such as brooded o'er the earth

When earth was chaos. The great Mastodon,

The dreaded monster of the elder world,

Passed o'er this mighty cavern, and his tread

Bent the old forest oaks like fragile reeds

And made earth tremble; armies in their pride

Perchance have met above it in a shock

Of war with shout and groan, and clarion blast,

And the hoarse echoes of the thunder gun;

The storm, the whirlwind, and the hurricane

Have roared above it, and the bursting cloud

Sent down its red and crashing thunder bolt;

Earthquakes have trampled o'er it in their wrath,

Rocking earth's surface as the storm wind rocks

The old Atlantic; yet no sound of these

E'er came down to the everlasting depths

Of these dark solitudes.

How oft we gaze

With awe or admiration on the new

And unfamiliar, but pass coldly by

The lovelier and the mightier! Wonderful

Is this world of darkness and of gloom,

But far more wonderful yon outer world

Lit by the glorious sun. These arches swell

Sublime in lone and dim magnificence,

But how sublimer God's blue canopy

Beleaguered with his burning cherubims

Keeping their watch eternal! Beautiful

Are all the thousand snow white gems that lie

In these mysterious chambers gleaming out

Amid the melancholy gloom, and wild

The rocky hills and cliffs, and gulf, but far

More beautiful and wild the things that greet

The wanderer in our world of light, the stars

Floating on high, like islands of the blest

The autumn sunsets glowing like the gate

Of far off Paradise; the gorgeous clouds

On which the glories of the earth and sky

Meet and commingle; earth's unnumbered flowers

All turning up their gentle eyes to heaven;

The birds with bright wings glancing to the sun,

Filling the air with rainbow miniatures;

The green old forest surging in the gale;

The everlasting mountains, on whose peaks

The setting sun burns like an altar flame;

And ocean, like a pure heart rendering back

Heaven's perfect image, or in his wild wrath

Heaving and tossing like the stormy breast

Of a chained giant in his agony.

GEOLOGY.

The following is an extract from an approximate section of the geological formations of the northern part of Edmonson county, Kentucky, by David Dale Owen, 1856:

"On the south of Green river, the platform of limestone forming the descent into the Mammoth Cave is 232 feet above Green river. The entrance to the cave being 38 feet lower than this bed of limestone, is 194 feet above Green river.

In the above 232 feet there are several heavy masses of sandstone, viz: at 125, 145, 150, 160 and 215 feet, but it is probable that most of them have tumbled from higher positions in the hill, as no alternations of sandstone have been observed at these levels in the cave.

From an elevation of 240 to 250 feet, the prevalent rock is sandstone without pebbles, which can be seen extending up to 312 feet to the foundation of the cave hotel.

The united thickness of the limestone bed on this part of Green river is about 230 feet, capped with 80 feet of sandstone.

About midway of the section on this part of Green river are limestones of an obscure oolitic structure, but no true oolite was observed.

Many of these limestones are of such a composition as to be acted on freely by the elements of the atmosphere, which in the form of nitric acid combine with the earthy and alkaline bases of calcareous rock, and give rise to the formation of nitrates, with the liberation of carbonic acid; hence the disintegrated rubbish of the caves yield nitrate of potash (saltpeter) after being treated with the ley from wood ashes and subsequent evaporation of the saline lixivium.

The wonderful cavernous character of the subcarboniferous limestones of the Green river valley, and, indeed, of these particular members of the subcarboniferous group, throughout a great part of its range in Kentucky and Indiana is due, in a great measure, to this cause, together with the solvent and eroding effect of water, charged with carbonic acid.

THE ROCK HOUSES

frequently encountered, both in this formation and in the limestones of Silurian date, are produced by similar causes; the more easily disintegrating, beds generally crumbling away while the more durable remain in the overhanging ledges.

HOW THE CRYSTALLINE FLOWERS ARE FORMED.

By the oxidation of other elements, sulphates of oxides of iron and alkalies, with carbonate of lime, give rise to the formation of gypsum, which appear in the form of rosettes, festoons and other imitative forms, on the walls and ceiling of the cave.

Crystallizations of sulphate of soda and sulphate of magnesia are not uncommon, both in some of the caves and in sheltered situations under the shelving rocks.

In one of the spacious halls beyond the Star Chamber, and through which the tourist must pass on his way to the Chief City, the sulphates of soda and magnesia in a very pure state are found in great quantities.

CONCLUSION.

We are frequently asked "At what season of the year can a visit to the Cave be most profitably made?" To this we answer, from our own experience we prefer the month of June, though from May till November, in ordinary seasons, the whole of the cave may be thoroughly explored. The most beautiful parts of the cave being beyond the rivers, can not be visited from November till May, as the streams in the cave are usually so swollen that no one ventures beyond them.

The rise and fall of these streams are known to be nearly fifty feet, filling up many of the passages altogether during a wet season. That part of the cave between the entrance and the rivers being much higher than the Green river, may be visited at any season, summer or winter. This is called the Short Journey, though the sum of its various avenues exceeds twenty miles in extent.

Every tourist will do well to put himself unreservedly under the care of his guide, and above all he should not become separated from the party, as serious consequences have been known to result from inattention to this important fact.

www.ingramcontent.com/pod-product-compliance
Lightning Source LLC
Chambersburg PA
CBHW070827180526
45168CB00002B/762